ENGLISH & AMERICAN FLOWERS

(1891)

BY

ALFRED RUSSEL WALLACE

British Library Cataloguing-in-Publication Data
A catalogue record for this book is available from the
British Library

Alfred Russel Wallace

Alfred Russel Wallace was born on 8th January 1823 in the village of Llanbadoc, in Monmouthshire, Wales.

At the age of five, Wallace's family moved to Hertford where he later enrolled at Hertford Grammar School. He was educated there until financial difficulties forced his family to withdraw him in 1836. He then boarded with his older brother John before becoming an apprentice to his eldest brother, William, a surveyor. He worked for William for six years until the business declined due to difficult economic conditions.

After a brief period of unemployment, he was hired as a master at the Collegiate School in Leicester to teach drawing, map-making, and surveying. During this time he met the entomologist Henry Bates who inspired Wallace to begin collecting insects. He and bates continued exchanging letters after Wallace left teaching to pursue his surveying career. They corresponded on prominent works of the time such as Charles Darwin's *The Voyage of the Beagle* (1839) and Robert Chamber's *Vestiges of the Natural History of Creation* (1844).

Wallace was inspired by the travelling naturalists of the day and decided to begin his exploration career collecting specimens in the Amazon rainforest. He explored the Rio Negra for four years, making notes on the peoples and languages he encountered as well as the geography, flora, and fauna. On his return voyage his ship, Helen, caught fire and he and the crew were stranded for ten days before being picked up by the Jordeson, a brig travelling from Cuba to London. All of his specimens aboard Helen had been lost.

After a brief stay in England he embarked on a journey to the Malay Archipelago (now Singapore, Malaysia, and Indonesia). During this eight year period he collected more than 126,000 specimens, several thousand of which represented new species to science. While travelling, Wallace refined his thoughts about evolution and in 1858 he outlined his theory of natural selection in an article he sent to Charles Darwin. This was published in the same year along with Darwin's own theory. Wallace eventually published an account of his travels *The Malay Archipelago* in 1869, and it became one of the most popular books of scientific exploration in the 19th century.

Upon his return to England, in 1862, Wallace became a staunch defender of Darwin's landmark work *On the Origin of Species* (1859). He wrote responses to those critical of the theory of natural selection, including 'Remarks on the Rev. S. Haughton's Paper on the Bee's Cell, And on the Origin of Species' (1863) and 'Creation by Law' (1867). The former of these was particularly pleasing to Darwin. Wallace also published important papers such as 'The Origin of Human Races and the Antiquity of Man Deduced from the Theory of 'Natural Selection" (1864) and books, including the much cited *Darwinism* (1889).

Wallace made a huge contribution to the natural sciences and he will continue to be remembered as one of the key figures in the development of evolutionary theory.

Wallace died on 7th November 1913 at the age of 90. He is buried in a small cemetery at Broadstone, Dorset, England.

ENGLISH AND AMERICAN FLOWERS
(1891)

I.

The numerous English writers who have described their impressions of North America tell us all about the people, their manners and customs, their hotels and churches, the mode of travelling and the scenery, the rivers and waterfalls, the mountains and forests, the prairies and deserts; but hardly ever do they give us any information as to the kind of vegetation that covers the surface of the soil, or the flowers that adorn the roadsides, the forests, or the mountains. Hence it comes to pass that the majority of English readers, even those who delight in the wild flowers of their own country or the more varied beauties of the Alpine flora, have usually the vaguest and most erroneous ideas as to what flowers are to be found in the United States and Canada, and to what extent they resemble or differ from those of our own country.

There are many circumstances which render it difficult, even for the native who is not a botanist, to learn much about American wild flowers. Confining ourselves at present to the North-Eastern States, we may say that three hundred years ago the whole country was covered with forest, and, with few exceptions, the herbaceous flowering plants were such as grew in the shade of trees or in the few open glades, in bogs, or on the banks of streams. Now, these forests have been so completely cleared away that comparatively little remains in its primitive condition, and often over extensive areas hardly a patch of original woodland is to be found. In other districts there is plenty of land covered with trees, but these are usually new growths, the timber having been felled again and again, as required for firewood, for fencing, or for other purposes. This wholesale clearing of the original forest-covering of the soil has led, no doubt, to the destruction of many lowly plants, some of which have become exterminated altogether, while others have been able to survive only in the few spots that still offer suitable conditions for their existence. Such places are comparatively rare, and often difficult of access; and hence the country, for a considerable distance round the larger cities and towns, affords but few of the really native plants, while common European weeds often abound. The old hedgerows, the shady banks and moist ditches, the deep-cut lanes, and the numerous footpaths of our own country, which afford abundant stations where wild flowers have been preserved to us from prehistoric times, are almost wholly wanting in America. There the seeker after wild flowers must usually be prepared to walk long distances over rough and pathless fields or hills in order to reach the places where alone he has any chance of finding the rarer or the more beautiful species. Owing to this absence of pleasant rural pathways the inhabitants of the towns rarely walk far into the country for exercise or pleasure unless they have some special pursuit of sport or natural history, and that want of interest in the natural productions of the district which is sufficiently common in England is still more prevalent in America.

The relations of the entire flora of temperate North America to that of Europe and Northern Asia have been the subject of much discussion among botanists. The late Professor Asa Gray made known, and, to some extent, popularised, the curious anomalies which these relations present, especially as regards the close affinity of the plants (more especially of the trees and shrubs) of the Eastern United States with those of Eastern Asia and Japan. Some account of Asa Gray's researches was given in this Review, in 1878, in an article on "Epping Forest," and they are only now referred to because they have been used to uphold the theory that, as regards the distribution of plants, the north temperate zone of the eastern and western hemispheres cannot be separated, but must be considered to form one botanical region. Recently, however, Sir Joseph Hooker has stated his opinion that if we go beyond the two fundamental botanical divisions--the tropical and temperate floras, which, for purposes of geographical distribution, are of little interest, we must consider that the temperate floras of the Old and

New Worlds are as distinct as are the tropical floras of the same areas; and he adds that, although the resemblances as regards certain genera and species of plants between Eastern America and Eastern Asia, is very remarkable, yet the temperate floras of Asia and America are in other respects totally dissimilar.[1] In the present article I shall endeavour to show, in a popular manner, what is the nature and extent of the dissimilarity between America and Europe as regards what are commonly known as wild flowers.

In order to restrict the inquiry within moderate limits, it is proposed to consider, first and mainly, the relations between the wild flowers of Great Britain as representing those of Western Europe, and those of Eastern North America as given in Asa Gray's *Botany of the Northern United States*, which includes the country from New England to Wisconsin, and southward to Ohio and Pennsylvania. This area may be fairly compared with that of England, France, and Germany, and will serve as a foundation for the wider comparison between North America east of the Mississippi with Europe, or of the whole of temperate North America with temperate Europe and Asia, to which occasional reference will have to be made. It must be understood that as our comparison regards only the native plants of the two countries, those numerous British or European species which have been introduced into America by human agency and have often become common weeds, will be left out of consideration altogether. We have to do only with the condition of the vegetation brought about by nature, undisturbed by the effects which have recently been produced by man.

There are two separate phenomena by which we may estimate the relations of the floras of two countries, both of which are important factors in the comparison--the absence from one country of whole groups of plants which are both common and widespread in the other, and the presence of new types entirely unknown in the other. It is usual to lay much more stress on the latter phenomenon, because the former occurs when there is no essential difference between the floras, the one having been recently derived from the other. Thus, many species, and even genera, of West European plants are absent from Britain, but this does not lead us to consider the British flora as being essentially different from that of Europe, the deficiencies being plainly attributable to the smaller area, the limited range of climate, the recent glacial epoch, and other such causes. But, when the country in which the deficiency occurs is fairly comparable with the other in all these respects, the cause of the phenomenon is evidently a deep-seated one, and must be held to show a fundamental diversity in their floras.

There are, of course, in every extensive flora such as that of North America a considerable number of almost cosmopolitan groups or species, and many others which are found in all temperate regions. Thus, no less than 115

European genera and 58 European species are found at the antipodes in New Zealand, and many others in Australia and South temperate America. Among these are such familiar plants as buttercups, anemones, poppies, violets, St. John's worts, gentians, forget-me-nots, many genera of cresses and other crucifers, mint, scull-cap, loose-strife, sea-lavender, and many others; and there are also in the same remote countries such common English species as the lady's-smock (*Cardamine pratensis*), chickweed (*Stellaria media*), the cut-leaved geranium (*Geranium dissectum*), the silver-weed (*Potentilla anserina*), the common bind-weed (*Calystegia sepium*), and scores of others, all considered to be indigenous and not introduced by man. It is evident, therefore, that we must expect to find a considerable number of English species in North America and a still larger number of English genera, because this is a feature which occurs in all temperate regions, and cannot be held to prove any *special* relationship between these two countries. Among these familiar English flowers we find a tolerable number of violets, anemones, St. John's worts, vetches, potentillas, willow-herbs, gentians, and some others; while wild geraniums, saxifrages, stonecrops, campanulas, forget-me-nots, and true orchises are far less frequently met with than with us.

But what most strikes the English botanist (next to the altogether unfamiliar types that everywhere abound) is the total absence or extreme rarity of many plants and groups of plants which are the most abundant and familiar of our native flowers, and which are almost equally common throughout Europe, and often throughout northern Asia. There are, for instance, no true poppies like those so abundant in our corn-fields, no common or musk-mallows of the genus Malva, or gorse or broom or rest-harrow, no teasel or scabious, no true heaths, no bugloss or comfrey, no ivy to adorn the old trees and walls with its glossy foliage, no mullein, toad-flax, snap-dragon, or foxglove, no scented thyme, basil or marjoram, no bright blue ground-ivy or bugle, no white or yellow or purple dead-nettles, no scarlet pimpernel, not even a primrose or a cowslip in all the land. There are, it is true, two species of Primula in the North-Eastern States, one the bird's-eye primrose of our northern counties, and another still smaller peculiar species, but both are confined to limited districts near the great lakes, and are not to be found unless specially searched for; and no other primroses are to be met with till we reach the Rocky Mountains, where there are two or three high alpine species.

Coming now to the endogenous plants, we find even more remarkable deficiencies. No daffodil, snowdrop, or snowflake is to be found wild in all North America, neither is there any crocus, wild hyacinth, colchicum, or lily-of-the-valley. The beautiful genus Ophrys, containing our bee, fly, and spider orchises, is quite unknown; and such familiar plants as the black-briony of our hedges, the flowering-rush

13

of our streams and ditches, and the curious butcher's broom of our dry woods, are nowhere to be met with.

Now the important thing to be noted is, that most of these plants are not only abundant and familiar in many parts of England but are widely spread throughout Europe, and the larger part of them belong to groups which extend into Northern Asia, and often reach the eastern extremity of that continent. If we take account of less important or less familiar plants this list might be doubled or trebled; and it might be still further extended if we took account of genera which range widely over Europe and Asia but happen to be rare or altogether wanting in England. Such, for example, are the following well-known garden flowers. The white and yellow asphodels, the red valerian (Centranthus), naturalised in many places on our chalk cliffs and old walls, the cinerarias, the gum-cistuses, the cyclamens, the daphnes, the true pinks (Dianthus), the numerous dwarf brooms (Genista), the corn-flags (Gladiolus), the candytufts (Iberis), the lavender and the rosemary, the ox-eye daisies, the stocks, the Star-of-Bethlehem, the pœonies, the mignonettes, the garden rue, the various soap-worts, the tulips, the periwinkles, and a hundred others.

It must always be remembered, that the British plants noticed above as being absent from the indigenous flora of the United States are abundant with us and form characteristic features of our flora, that the larger portion of them range widely over Europe and Western Asia, that more than half of them extend across Northern Asia to the Pacific and often to China and Japan, while several extend over the greater portion of the eastern hemisphere, and are found also in Australia or South Africa. The absence of such a number of the characteristic and dominant groups of plants of the temperate zone from so extensive and varied an area as the United States and Canada, is of itself a very remarkable phenomenon, and affords a *primâ facie* ground for treating the temperate regions of the New World as a distinct botanical region.

Another feature to which botanists attach much importance in the comparison of separate floras is the proportionate abundance of the various orders and tribes in the two countries, which, when very different, leads to the general vegetation having a distinctive aspect. In this respect, Europe and Eastern America differ greatly. Among the most abundant and characteristic groups, which everyone recognises in our own country and in Europe as common plants everywhere to be met with, are those of the cabbage and cress tribe (Cruciferæ), the pink family (Caryophyllaceæ), the umbel-bearers (Umbelliferæ), the thistle-tribe of the Compositæ, the bluebells (Campanulaceæ), the primroses (Primulaceæ), and the orchises (Orchidacæ); but all these are much less frequent in North America, and are usually so scarce as to take little or no part in determining the special character of the vegetation. As an illustration of the difference, there are only twelve indigenous genera of Cruciferæ in the North-Eastern United States with about thirty-five indigenous species, while the comparatively poor British flora possesses twenty-four genera, and fifty-four species.

Instead of these characteristic European types we have in America some peculiar Rubiaceæ, among which is the pretty creeping Mitchella or partridge berry, and an enormous preponderance of Compositæ, including numbers of non-European genera and a great variety of eupatoriums, asters, golden-rods, and sun-flowers, together with some of our well-known garden flowers such as Liatris, Rudbeckia and Coreopsis. The milk-worts (Polygala) are rather numerous, and the milk-weeds (Asclepias) still more so, and these last are quite unlike any European plants. The beautiful phloxes are a very characteristic type almost exclusively confined to North America, and often affording brilliant patches of floral colours. Among the endogenous plants the numerous specie of Smilax, well called "green-briar," are highly characteristic and peculiar, though the genus is found in Southern Europe, while the beautiful wood-lilies of the genus Trillium (found again in Eastern Asia) have curious or ornamental flowers. Add to these the interesting spider-worts forming the genera Commelyna and Tradescantia, and having their allies in the tropics, and we have completed the enumeration of the more conspicuous groups of non-European herbaceous flowering plants which give a distinctive character to the flora.

There are, however, many other plants which, although belonging to small orders and not represented by more than one or a very few species, are yet so abundant in individuals, and so widely spread over the country, as to contribute largely to the general impression which the North American flora produces on an English botanist on his first visit to the country. This may be illustrated by a brief account of some of the present writer's rambles in search of American flowers.

My first walk was on the 13th February, in the woody country to the north-west of Washington. Here we found on dry banks the beautiful little May-flower (*Epigæa repens*), and the pretty spotted wintergreen (*Chimaphila maculata*), both members of the heath family, and both of genera almost peculiar to America, allied species of each being found in Japan, while some of the forest slopes were covered with the beautiful *Kalmia latifolia*, another peculiarly American genus of Ericaceæ. The curious parasitic "beech-drops," one of the Orobanche tribe, is also peculiar, while the "partridge-berry" (*Mitchella repens*) has its only ally in Japan. Other plants observed were the *Aralia spinosa* or Hercules' Club, a small tree of a non-European genus, a species of Smilax like a slender leafless rosebush, a hairy-leaved blackberry (*Rubus villosus*), a wild vine, a white hepatica in flower identical with the European species, the curious skunk-cabbage (*Symplocarpus fœtidus*), belonging to the Arum family and also in flower, the "blazing-star" (*Chamælirium luteum*), one of the colchicum tribe and peculiar to North America, the rattle-snake plantain (*Goodyera pubescens*), really an orchis, of which there is one European species found also in Scotland, and a handsome sedge, *Carex platyphylla*. Thus, out of fourteen plants distinguished at this wintry season, only three belonged to British, and four or five to European genera, while the large majority were either quite peculiar to America or only found elsewhere in Japan, Eastern Asia, or the Pacific Islands. During another excursion to the rich locality of High Island, five miles from Washington, on March 27th, several other interesting and characteristic plants were found. Such were the "spring beauty" (*Claytonia virginica*), a pretty little rosy white flower belonging to the Portulaceæ which often carpets the woods and takes the place of our wood-anemone, for though there are several wood-anemones in America they do not form so important a feature of the spring vegetation as with us. The rare and pretty "harbinger of spring" (*Erigenia bulbosa*), a minute umbelliferous plant quite peculiar to America, grew here abundantly, as did the pinnate-leaved Virginian water-leaf (*Hydrophyllum virginianum*). The may-apple (*Podophyllum peltatum*), and the twin-leaf (*Jeffersonia diphylla*), herbaceous plants of the Berberis family, were abundant, the former occurring elsewhere only in the Himalayas, while the latter is North American and East Asiatic. A yellow violet, a perennial chickweed, a tooth-wort (Dentaria), a stone-crop, and an unconspicuous saxifrage, alone showed any resemblance to our native vegetation.

About the middle of April, in the vicinity of Cincinnati, I was introduced to the spring flowers of the North-Eastern States, in their full development. The woods were here carpeted in places with the "spring beauty," while in other parts there were sheets of the curious "Dutchman's breeches" (*Dicentra cucullaria*), like a small yellow dielytra, to which it is allied. Then there were smaller patches of the *Thalictrum anemonoides*, resembling a very slender wood-anemone, the exquisite little "Blue-eyed Mary," sometimes called "Innocence," (*Collinsia verna*), the handsome celandine-poppy (*Stylophorum diphyllum*), like our "greater celandine," but with larger and more richly coloured flowers, the elegant lilac-coloured *Phlox divaricata*, and the "blood-root" (*Sanguinaria canadensis*), with its beautiful white star-shaped flowers.

Here, too, the buds of the handsome purple wood-lily (*Trillium erectum*) were just showing themselves, and there were large patches of the yellow and white American dog's-tooth violet (*Erythronium Americanum*), just coming into bloom. In a damp river-bottom, the exquisite blue *Mertensia virginica* was found. It is called here the "Virginian cowslip," its drooping porcelain-blue bells being somewhat of the size and form of those of the true cowslip, but the plant is really allied to our lungworts. More homely-looking plants were a creeping yellow buttercup, with blue, white, and yellow-flowered violets, but they were utterly insignificant as compared with the many new and strange forms that constituted the bulk of the vegetation.

At the end of July I had the opportunity of seeing the swampy forests of Michigan, with their abundance of ferns, their pitcher plants (Sarracenia), yellow-fringed orchises (*Habenaria ciliaris*), and the curious little gold-thread (*Coptis trifolia*), found also in Arctic Europe, and so named from its yellow thread-like roots,--all three growing in the dense carpet of sphagnum moss which covers the ground to the depth of one or two feet. In the cleared marshy ground, and along the margins of the streams and ditches, was a dense vegetation of asters, golden-rods, and other composites, many of which were of groups unknown in Britain or in Europe, while still lingering on the burnt-up road sides were the handsome flowering spurge (*Euphorbia corollata*), with its curious white flowers, and the elegant foliage of the bird's-foot violet.

A few remarks on the general aspects of the country as regards native vegetation and flowers must conclude this very imperfect sketch. What most impresses the nature-loving Englishman while travelling in America is, the newness and rawness of the country, and the almost universal absence of that harmonious interblending of wild nature with human cultivation, which is so charming over a large part of England. In these North-Eastern States, the native forests have been so ruthlessly destroyed, that fine trees are comparatively rare, and such noble elms, beeches, oaks, and sycamores as are to be found arching over the lanes and shading the farmhouses and cottages in a thousand English villages, are only to be seen near a few of the towns in the older settled States, or as isolated specimens which are regarded as something remarkable. Instead of the old hedge-rows with tall elms, spreading oaks, and an occasional beech, hornbeam, birch, or holly, we see everywhere the ugly snake-fence of split rails, or the still more unsightly boundary of barbed wire. Owing to the country being mostly cut up into one-mile square sections, subdivided into quarters, along the outer boundaries of which only is there is any right-of-way for access to the different farms, the chief country roads or tracks zig-zag along these section-lines without any regard to the *contours* of the land. It is probably owing to the cost of labour and the necessity of bringing large areas under cultivation as quickly as possible, that our system of fencing by live hedges, growing on a bank, with a ditch on one side for drainage, seems to be absolutely unknown in America; and hence the constant references of English writers on rural scenery and customs to "the ditch," or "the hedge," are unintelligible to most Americans.

The extreme rapidity with which the land has been cleared of its original forest seems to have favoured the spread of imported weeds, many of which are specially adapted to seize upon and monopolise newly exposed or loosened soil; and this has prevented the native plants, which might have adapted themselves to the new conditions had the change gone on very slowly, from gaining a footing. Hence it is that the cultivated fields and the artificial pastures are less flowery than our hedge-bordered fields and old pastures, while the railway banks never exhibit such displays of floral beauty as they often do with us. An American writer in *The Century* for June, 1887, summarises the general result of these varied causes, with a severe truthfulness that would hardly be courteous in a stranger, in the following words:--

"A whole huge continent has been so touched by human hands, that over a large part of its surface it has been reduced to a state of unkempt, sordid ugliness; and it can be brought back into a state of beauty only by further touches of the same hands more intelligently applied."

Let us hope that intelligence of this kind will soon be cultivated as an essential part of education in all American schools. This alone will, however, have no effect so long as the fierce competition of great capitalists, farmers and manufacturers, reduces the actual cultivator of the soil, whether owner, tenant, or labourer, to a condition of sordid poverty, and a life of grinding labour which leaves neither leisure nor desire for the creation or preservation of natural beauty in his surroundings.[2]

Although with the limited opportunities afforded by one spring and summer spent in America, it is impossible to speak with certainty, yet both from my own observation, and from information received from residents in various parts of the Eastern States, it seems to me, that in no part of America, east of the Mississippi, is there such a succession of floral beauty and display of exquisite colour as are to be found in many parts of England. Such, for instance, are the woods and fields of daffodils, "which come before the swallow dares, and take the winds of March with beauty"; the wild hyacinths, whose nodding bells, of exquisite form and colour individually, carpet our woods in April with sheets of the purest azure; the soft yellow of primroses in coppices or along sunny hedge banks; the rich golden yellow of the gorse-bushes which, when seen in perfection as in the Isle of Wight, Cornwall, or Ireland, is so superlatively glorious, that we cannot wonder at the enthusiasm of the great Linnæus, who, on beholding it, knelt down and thanked God for so much beauty; later on the clearer yellow of the broom is hardly less brilliant on our heaths and railway banks, while the red ragged-robin, and the purple or rosy orchises often adorn our marshes and meadows with masses of colour; then come the fields and dry slopes, gay with scarlet poppies, and the noble spikes of foxgloves in the copses and on rough banks, followed by, perhaps, the most exquisitely beautiful sight of all, the brilliant sheets and patches of purple heath, sometimes alternating with the tender green of the young bracken, as on some of the mountain slopes in Wales, sometimes intermingled with the rich golden clumps of the dwarf gorse, as on the wild heaths of Surrey or Dorset.

Truly, the Englishman has no need to go abroad to revel in the beauty of colour as produced by flowers. Although the number of species of plants which inhabit our islands is far less than in most continental areas of equal extent, although the gloom and grey of our skies is proverbial, and we want the bright sunshine of American or Eastern summers, yet these deficiencies do not appear to lessen the luxuriant display of bright colours in our native plants. The mountains of Switzerland, the arid plains of the Cape and of Australia, the forests and swamps of North America, provide us with thousands of beautiful flowers for the adornment of our gardens and greenhouses, yet, from the descriptions of these countries by travellers or by residents, it does not seem that any one of them produces a succession of floral pictures to surpass, or even to equal, those which the changing seasons display before us at our very doors. The absence of fierce, long-continued sunshine, which renders it difficult for us to grow many fruits and flowers which flourish even in the short Canadian summer, lengthens out our seasons favourable to vegetation, so that from the violets and daffodils of March, to the heaths and campanulas, the knapweeds, and the scabious of September or October, we are never without some added charm to our country walks if we choose to search out the appropriate spots where the flowers of each month add their bright colours to the landscape.

To the botanist, the poverty of our English flora contrasts unfavourably with the number of species and the strange or beautiful forms to be found in many other temperate regions, and to him it is a great delight to make the acquaintance, for the first time, in their native wilds, of the many curious plants which he has only known before in gardens or in herbaria. But the simple lover of flowers, both for their individual beauty and for the charm of colour they add to the landscape, may rest assured that, perhaps with the single exception of Switzerland, few temperate countries can equal, while none can very much surpass his own.

II.

FLOWERS AND FORESTS OF THE FAR WEST.

Temperate North America, as regards its types of vegetation, consists of four well-marked subdivisions. The most important and the richest in species is the great forest region of the Eastern States, whose main peculiarities were indicated in the first part of this article. West of this area, and extending from a short distance beyond the Mississippi to the base of the Rocky Mountains is the region of the great plains, almost destitute of trees, except in the river bottoms, but with a fairly rich herbaceous flora; and a very similar vegetation is found in the half-desert valleys and plains between the Rocky Mountains and the Sierra Nevada. A third botanical district consists of the higher wooded portions of the Rocky Mountains, together with the peaks and high valleys above the timber-line, in which the vegetation is, in many respects, very distinct from that of any other part of temperate America. Lastly comes the Californian region, extending from the Pacific coast to the upper limit of trees in the Sierra Nevada, a country of surpassing interest to the botanist, and well-known to every lover of flowers for the great number of beautiful and peculiar forms it has furnished to our gardens. It is proposed to give a brief sketch of the more prominent features of the flora of the three western regions, derived partly from personal observation during a summer spent in the country, largely supplemented by the writings of the late Professor Asa Gray and other American and English botanists.

The first region to be considered, that of the prairies, the great plains, and the deserts of the inland basin, may be very briefly noticed, since, although of considerable interest to the botanist, it is only occasionally that plants, remarkable for beauty of flower or other conspicuous characteristics, are met with. The eastern portion of the district, where the rich prairie lands of Kansas and Nebraska are being rapidly cultivated, produces many fine flowering plants wherever some steep or rocky slope has escaped cultivation. Here we find abundance of yuccas intermingled with blue pink and white-flowered spider-worts, handsome large-flowered penstemons, baptisias with large pea-like flowers of blue, yellow or white, many species of astragalus, yellow and white evening-primroses and other allied forms, several cactuses of the genera opuntia and mammillaria, blue larkspurs, pink oxalises, the purple Phlox divaricata, mallows of the genera Malvastrum and Callirhoe, some of which are well-known garden plants, and a host of sunflowers, asters, cone- flowers, golden-rods, coreopsis, and many other showy composites.

This is the region of the buffalo or bunch grasses which formed the chief subsistence of the American bison. They are fine-tufted bluish grasses, much resembling in appearance our fine-leaved bent grass (*Agrostis setacea*), which is common on the heaths about Bournemouth and in Dorsetshire. I was informed that since the bisons had been destroyed the buffalo grass was also disappearing, being replaced by various coarser growing plants and grasses. It is probable that the uniform hardening of the surface by the tread of the herds of bison, together with the equally regular manuring, favoured the growth of this particular form of grasses.

As we travel westwards, towards the Rocky Mountains, the plains become more arid, and in places the vegetation resembles that of the deserts of the great basin. Here there are fewer conspicuous flowers, and a preponderance of dwarf creeping plants, with a few thorny bushes and some species of wormwood, forming the well-known "sage-brush" of the deserts. In the interior plains these thorny and grey-leaved shrubs prevail, with wide tracts of bare earth often covered with saline incrustations. Here and there are found some pretty flowers, such as phloxes, alliums, phacelias, gilias, cleomes, onotheras, and other characteristic plants; but the general aspect is that of bare soil scantily covered with a dwarf vegetation, or of low, shrubby thickets of a grey or leafless aspect, consisting mostly of plants allied to the salt-wort, orache, and sea-blite of our salt marshes, or the goosefoot and wormwood of our waste places.

We will now leave these comparatively uninteresting plains and deserts and enter on the Rocky Mountains proper, their deep cañons, their wooded slopes and valleys, and their upland pastures, rocky streams and alpine heights. The forest trees consist mainly of a few species of pines, firs, and junipers, none of them very remarkable for size or beauty, with several poplars, and a few oaks, beeches, and maples; but these rarely form continuous forests, except where the soil and other conditions are especially favourable. Almost everywhere the conifers are most prominent, and give their peculiar character of dark ever-green spiriness to the forest vegetation. The present scantiness of timber trees is no doubt partly due to the agency of man, first by starting forest fires, which rapidly clear extensive areas, and more recently by the felling of timber for building and mining, a cause which has denuded most of the valleys of their original forest trees. There are a considerable number of shrubs of the usual American types, such as sumachs, snowberries, hazels, spiræas, brambles, and roses, mostly of species common to other parts of America and of no special interest from our present point of view.

It is when we enter among the mountains and explore the valleys, cañons, and lower slopes, that we meet with a variety of new and interesting plants. Among these are some which are specially characteristic of this part of North America. The phloxes, polemoniums, and gilias, some species of which are common in our gardens, are abundant, as are the penstemons and mimuluses, with the brilliant castilleias belonging to the same family (Scrophulariaceæ), whose crimson or scarlet bracts form one of the greatest ornaments of the higher woods and pastures. The elegant genus Phacelia is not uncommon, though its chief development is in California, and the moist valley-bottoms are often blue with the well-known flowers of the bulbous camassia. A curious genus of the Polygonum family (Eriogonum) is abundant, and the yellowish-white or rosy flowers of some of the species are very pleasing. Handsome composites abound, especially the genus Erigeron, with a number of peculiar forms, while the beautiful butterfly-tulips of California here make their first appearance. Lupines also are plentiful, though less so than further west, and the beautiful American cowslips (Dodecatheon) sometimes called "shooting-stars" are not unfrequent in boggy meadows.

But in addition to these more or less characteristic American types, the botanist is at once struck by the appearance of a number of European or even of British plants, and these not introduced weeds but forming an essential part of the flora. This is proved by the fact that the further we penetrate among the mountains and the higher we ascend, the more numerous become these familiar species or genera. Among the more abundant of these plants are the common yarrow (*Achillea millefolium*), our blue hare-bell (*Campanula rotundifolia*), the bistort (*Polyonum bistorta*), the common silver-weed of our roadsides (*Potentilla anserina*), and the rarer shrubby cinquefoil (*P. fruticosa*). In the sub-alpine and alpine districts these plants of the old world become more frequent and occupy a larger space in the entire vegetation, and in order to show the importance of this interesting feature of the Rocky Mountain flora it may be well to give a brief account of a week's exploration of the vicinity of Gray's Peak, one of the highest mountains of Colorado.

Accompanied by a botanical friend from Denver I went first by rail up Clear Creek Cañon, passing by Georgetown, to Graymount, the terminus of the railway, where there is a hotel and where horses are obtained for the ascent of Gray's Peak, about nine miles distant by the road. Graymount is situated at the junction of two valleys and is about 9,500 feet above the sea level. During a short stroll on the afternoon of our arrival on some rocky slopes we found two of our rarer British plants, the winter green (*Pyrola rotundifolia*) and the elegant twin-flower (*Linnæa borealis*), but instead of having nearly white flowers the former was reddish and the latter was of a deeper colour than in our native plant. The next day we walked to Kelso's cabin, where are some miners' houses about 11,000 feet above sea-level, situated at the lower end of a fine upland valley, which is above the timber line. During the earlier part of our walk up a wooded

31

valley we first noticed fine clumps of the Siberian lungwort with its lovely pale blue flowers, growing more compactly than in our gardens, and splendid masses of the shrubby cinquefoil covered with its handsome yellow flowers, as well as our common harebell, all in the greatest luxuriance and beauty. In damp shady places we found the little moschatel, and in bogs the curious Swertia perennis, a kind of gentian with slaty-blue flowers. These are all European or North Asian plants, but there were many others peculiar to the region though sometimes of European rather than American affinity. Such are the lovely columbine (*Aquilegia cærulea*), allied to the species of the European Alps, abundant and conspicuous with its large blue and white flowers, while mingled with it grew the gaudy Castilleia integra, whose leafy bracts of intense crimson are visible from a long distance. This is a true American type, as is the pretty liliaceous plant, Zygadenus glaucus; and there were also abundance of dark purple or bright blue penstemons, several showy groundsels and erigerons and the handsome yellow composite, Arnica cordifolia.

It was when we had passed the timber line at about 11,500 feet elevation, and had entered the bare rocky valley at the head of which rises the snow-flecked summit of Gray's Peak, that we discovered some of the chief gems of the alpine flora of the Rocky Mountains. Along the borders of the stream, fed by the still melting snows and with its roots in the water, were fine clumps of the handsomest American primrose (*Primula Parryi*), its whorled flowers of a crimson-purple colour with a yellow eye resembling in general appearance the well-known Japanese primrose of our gardens. Among the stony *débris* and loose boulders which bordered the stream the beautiful Phacelia sericea was abundant, its violet-blue flowers growing in dense clusters and producing a charming effect among its desert surroundings. This is a typical American plant, since not only is the genus a peculiar one but the natural order to which it belongs--the Hydrophyllaceæ--is almost confined to that continent. The beautiful nemophilas of our gardens belong to the same family. In boggy places the handsome Greenland lousewort, an Arctic species, was plentiful, and in rocky crevices we found the moss campion (*Silene acaulis*), which is abundant on the Scotch and Welsh mountains.

The next morning we fortunately determined to explore a lateral valley called Grizzly Gulch, which diverged to the north a mile above the hotel and led into a fine upland valley on the north side of Gray's Peak. Here, just below the timber-line, we found a miner's house, and the two miners who had come home to dinner invited us to join them, and then offered to show us a fine place for flowers. They took us through the wood for half a mile, when we came upon a rocky and grassy slope with great snow-patches in the shady hollows, and the ground which the snow had left was literally starred with flowers. Leaving us to go to their work

in a mine on the steep side of the mountain, we luxuriated
in the finest Alpine flower-garden we had yet seen, although
my friend had visited the mountains several times. What
first attracted our notice were three plants of the crowsfoot
family, which grew intermingled on a grassy slope almost
surrounded by snow. These were, a nearly white globe-
flower (*Trollius albifloras*), very dwarf and with spreading,
not globular flowers; a buttercup, whose flowers were of the
most perfect circular outline, and of a pure and rich yellow,
both peculiar to the Rocky Mountains; and the narcissus-
flowered anemone of the European Alps. Going a little
further we found some of the more characteristic American
forms, such as the beautiful blue-flowered Mertensia alpina,
a dwarf Alpine form of Mertensia siberica and perhaps the
most lovely plant of the genus; the pretty fringed grass
of parnassus (*Parnassia fimbriata*); with peculiar species
of the European genera, Aster, Cardamine, Astragalus,
Delphinium, Trifolium, Saxifraga, Sedum, Valeriana,
Veronica, and Pedicularis; with others of the American
genera, Phacelia, Chionophila, Mimulus, and Zygadenus;
and hidden among the rocks the minute purple-flowered
Primula augustifolia. What more especially interested me,
however, was the number of identical British or European
species. Such were the moss-campion, the Dryas octopetala,
Sibbaldia procumbens, the rosewort (*Sedum rhodiola*),
the Alpine Veronica, and the Alpine chickweed, Lloydia

serotina, a small liliaceous plant found on Snowdon, and two saxifrages, Saxifraga nivalis and S. cernua, all found also in our Welsh or Scotch mountains; and of European Alpines the pretty slaty-blue Swertia perennis which dotted the grassy slopes with its delicate flowers, the Alpine Astragalus, the Arctic willow, several saxifrages and gentians, and some other species characteristic of the flora of the Alps.

The next day, after sleeping at a miner's cabin situated at the head of the main valley at about 12,500 feet elevation, we ascended to the top of Gray's Peak, which is 14,250 feet high, and met with many other interesting plants. The little Eritrichium nanum, a minute but intensely blue forget-me-not, was found growing in the midst of clumps of the moss-campion; the Gentiana tenella and Campanula uniflora of the Arctic regions were also found at about 13,000 feet elevation; with the British Alpine penny-cress, the yellow Iceland poppy, the two-flowered sandwort, the Alpine arnica, the snowy buttercup, and other truly Arctic plants. Along with these were a few American alpine types, such as Eriogonums, Castilleias, and several composites. Near the summit of the mountain there were alternate upward-sloping bands of loose rock-débris and short turf, the latter gay with pretty yellow flowers. On examination these were found to consist of a potentilla and a saxifrage, whose flowers, resting close on the ground, were so much alike in form and colour that at a short distance they appeared identical. The intermixture of two very distinct species of flowers, coloured and shaped alike and flowering at the same time, is very uncommon, because it would interfere with regular cross-fertilization by insects. In this high and exposed situation, however, where the flowering season is very short and insects very scarce, the combination of two species of flowers may lead to a more conspicuous display, and be more attractive to whatever insects may visit such great altitudes; while with plants of such distinct families, the intermixture of the pollen would lead to no evil result, since each would be totally inert on the stigma of a flower of the other kind. The two species appear to be Saxifraga chrysantha and Potentilla dissecta.

On a general summary of the plants noticed during this excursion to one of the richest districts in the Rocky Mountains, I find that they comprised no less than 20 British species, about 45 European, mostly high Alpine or Arctic, and about 30 species which, though distinct, were allied to European types. There were thus a total of 95 species, either identical with or allied to European plants, while those which belonged to American genera, or were most nearly allied to American species, were about 30 in number. It thus appears that the alpine flora of the Rocky Mountains is mainly identical with that of the Arctic regions, and it is this identity which leads to the occurrence of so many British species in this remote district. In the review of the entire alpine flora of the Rocky Mountains by Professor Asa Gray and Sir Joseph Hooker, the number of species identical with those of the Arctic regions is 102, and the distinct, though often allied, species 81, while those that belong to peculiar American genera are only 14 in number.

In considering how this curious similarity of the alpine species of the two continents has been brought about, we must go back to a time anterior to the glacial epoch, when a rather mild climate prevailed in much of what is now the Arctic regions. The present Arctic flora, or its immediate ancestors, was then probably confined to the highest latitudes around the North Pole, together with the higher mountains which were immediately contiguous--such as Greenland, then only partially or not at all ice-clad, Spitzbergen and Nova Zembla, and some of the mountain peaks of Alaska and North-Eastern Asia. At this time the Rocky Mountains, the European Alps, and even Scandinavia supported in all probability only alpine forms of the plants of the surrounding lowlands, such as are now everywhere intermingled with the widespread Arctic species. As the cold came on, and the ice sheet crept farther and farther over the two continents, the true Arctic plants were driven southward, displacing the indigenous flora, which could not withstand the increasing severity of the climate, and occupying all the great mountain ranges on the lower side of the ice-fields and glaciers, and also such of the peaks as rose permanently above the ice-sheet of the glacial epoch. As the cold period gradually passed away, these hardy plants kept close to the gradually retreating ice, and in this way mounted to the higher peaks of many mountains from which the ice and even perpetual snow wholly passed away. Thus it is that so many species are now common to the Rocky Mountains and the European Alps; and, what seems more extraordinary, that identical plants occur on the summits of the isolated Scotch and Welsh mountains, and also on the White Mountains of New Hampshire and some of the mountains to the south of them.

Before passing on to sketch the flora of the west coast of America, we may briefly notice the more prominent differences between the Rocky Mountain flora and that of our European Alps, such differences as must strike every traveller who takes an interest in the floral beauties of the two regions. In the Alps the more striking and showy flowers of the Alpine pastures and higher rocks are the white, purple, and yellow anemones; the beautiful violas; the glorious blue gentians starring the short turf with azure and indigo, the numerous saxifrages, often with large and showy sprays of flowers; the many beautiful rosy and purple primulas and yellow auriculas; the handsome pinks; the delicate campanulas; the showy white and yellow buttercups, and the graceful meadow-rues. Now in almost all these groups the Rocky Mountain alpine and sub-alpine flora is deficient. Anemones are comparatively few in species and not abundant; violas are almost absent in the higher regions; gentians, though fairly abundant in species, make no brilliant display as they do in the Alps; saxifrages are few, and those of the crusted section with rigid leaves and large racemes of flowers are entirely wanting; primulas are represented by one handsome and two small and rather scarce species; campanulas are scarce, and pinks are entirely absent; while buttercups and meadow-rues are by no means abundant. Instead of these flowers so familiar to the Alpine tourist, the most showy and widespread plants are the fine long-spurred blue and white columbine, and the scarlet or crimson-bracted castilleias, which form sheets of beautifully contrasted colours, often covering wide mountain slopes either above or just below the timber-line; numerous purple or blue penstemons; fine blue polemoniums and lungworts of the genus Mertensia; some handsome purple or whitish louseworts, and a host of showy purple or yellow composites, which are far more numerous and varied than in the European Alps, and occupy a more prominent place in the alpine and especially in the sub-alpine Rocky Mountain flora. It is evident, therefore, that, notwithstanding the identity of so many of the species and genera of the two regions the proportions in which they occur are very different, and the aspect of the two floras is thus altogether distinct, and in some respects strikingly contrasted.

When we go westward to the Sierra Nevada of California, we meet with another alpine flora, generally similar to that of the Rocky Mountains, but with a smaller proportion of Arctic species and more which are characteristic of America. Here we find dwarf shrubby penstemons, curious prickly gilias, Mimulus and Eriogonum in more abundance, and a greater variety of ferns. But it is when we descend to the lower slopes and to the valleys and coast ranges of California itself that we find the greatest abundance of new plants altogether distinct from anything in the Eastern States, and it is to these that we must devote the remainder of our space.

Few countries have contributed to our gardens a larger number of showy and interesting plants than California. The rich orange yellow Eschscholtzias, the brilliant Calandrinias, the showy Godetias and Clarkias, the beautiful little Nemophilas and Phacelias, the gaudy Mimuluses and the handsome Collinsias, are known to every lover of garden flowers. Others familiar to every horticulturist are the curious pitcher-plant--Darlingtonia Californica, the handsome gigantic white poppy--Romneya coulteri, the elegant Dicentra formosa, the fine yellow-flowered shrub Fremontia Californica, the ornamental blue or white flowered evergreens of the genus Ceanothus, the fine shrubby lupines, the lovely flowering currants, including the fine Ribes speciosum with drooping fuchsia-like flowers, the scarlet-flowered Zauschneria Californica, the fine shrubby Diplacus glutinosus, and lastly, the many ornamental bulbous plants, such as the triteleias, brodiæas, lilies, and especially the lovely butterfly-tulips of the genus Calochortus, whose flowers are most exquisitely marked inside with delicately-coloured hairy fringes. But this by no means gives an idea of the great peculiarity of the Californian flora, which is best shown by the large number of its genera, probably more than a hundred, which are altogether unknown in the Eastern States. The flora is in fact related to that of Mexico, just as the flora of the Rocky Mountains is related to that of the Arctic regions, and the Eastern States flora to that of Japan and Eastern Asia.

But although the valleys and lowlands of California are specially characterised by hosts of brilliant annuals, monkey-flowers, lupines, and flowering shrubs, which make the country a veritable flower-garden in early spring, it is from its mountain forests of coniferæ that it derives its grandest and best-known characteristics. To a brief sketch of these, and of the accompanying shrubby and herbaceous vegetation, the remainder of this article will be devoted.

The Sierra Nevada of California, though rising to nearly the same altitudes as the Rocky Mountains, is by no means an imposing range, owing to the exceedingly gradual slope of the foothills which are continuous with it. From these low and arid hills, rising with a very moderate slope from the great central valley of California, there is a constant rise over an undulating or rugged country for nearly a hundred miles to the summits of the great range. The intervening tract is often cut into deep winding valleys, whose higher slopes are terminated by rugged volcanic precipices, where they have cut through the old lava-streams that once covered a large portion of the mountains; while nearer to the crest are enormously deep valleys, bounded with vertical walls and gigantic domes or splintered peaks of granitic rocks, of which the celebrated Yosemite Valley is the best known example. Owing to this formation the summits of the range can only be seen from great distances and from a few favourable points, as a somewhat jagged line on the far horizon, just rising above the dark forest-clad slopes, and here and there flecked with perpetual snows. A coach drive of three days from the railway terminus at Milton to the Yosemite Valley, and another to the Calaveras groves of "big trees," gave me an excellent opportunity of observing the main features of this remarkable forest region.

The lower portion of the foothills up to two or three thousand feet has been greatly defaced by gold-miners, who have dug over miles of ground and cleared away most of the fine timber. This lower portion is, however, naturally more arid, and the trees have never been so fine as at greater elevations. It is curious to notice how the pines and firs increase in beauty as well as in size as we ascend further towards the central ranges. For the first thousand feet there is a scanty vegetation of stunted shrubs, and the only conifer is the scrub-pine (*Pinus sabiniana*) which has a most singular appearance, being irregularly branched, with scanty foliage, and when well grown, looking at a distance more like a poplar than a pine. Higher up occurs the large white pine (*Pinus ponderosa*), which, except in very fine specimens, is a coarse, unornamental tree. Above two thousand feet we meet with the sugar pine (*Pinus lambertiana*), so called because its turpentine is sweet and sometimes almost like a mixture of sugar and turpentine. This is a handsomer species, and when full grown is of immense size and may be known at a distance by its clusters of large cones hanging down from the very extremities of its loftiest branches. Thus far the forests are poor, owing to the absence of the more elegant firs and cedars which only appear above 2,500 feet, when we first meet with the noble Douglas fir and the beautiful Red cedar (*Libocedrus decurrens*). This last is usually known in our gardens as Thuja gigantea, characterised by its columnar mode of growth and here sometimes reaching a hundred and fifty feet in height. Higher still, at about 4,000 feet, we come upon the most beautiful of the Californian firs, *Abies concolor* and *A. nobilis*. Both are exquisitely symmetrical in growth, while the dense horizontal branches of the latter species are adorned with the most delicate blue-green tints. These beautiful trees are to be seen here in every stage of growth, from such small plants as we see on the lawn of a suburban villa up to noble specimens 150 or 200 feet in height. These two elegant firs, along with the stately cedar and Douglas fir, and the noble yellow pine and sugar-pine, constitute the main bulk of the forest from 4,000 to 7,000 feet elevation, the belt in which alone are found the true "big trees" (*Sequoia gigantea*), in this country commonly known as the Wellingtonia.

Throughout these magnificent forests there is hardly any admixture of exogenous trees, and those that do occur only form an undergrowth to the far loftier coniferæ. A few small oaks and maples are sometimes seen, but more generally there is only an undergrowth of beautiful shrubs, the most conspicuous being the fine Californian dogwood, whose flowers, formed of the white involucres, are six inches across; and the lovely white azalea, whose delicate blossoms are beautifully marked with yellow. Besides these are the handsome Californian laurel and the white or blue flowered Ceanothus, while the "madrono" and "manzanita" (species of Arbutus and Arctostaphylos), are found in the drier portions of the forest and at a lower elevation.

The ground under the pines and firs is usually rather bare, but in favourable places there are some curious or beautiful creeping or herbaceous plants. Some of the drier slopes are completely carpeted with a curious little rosaceous plant (*Chamæbatia foliolosa*), having white flowers like those of a bramble and the most minutely divided tripinnate foliage, each leaflet looking about the size of a pin's head. Perhaps the most remarkable herbaceous plant of these forests is the

Sarcodes sauguinea, a leafless parasite allied to our native monotropa, but of an intense crimson colour and very large, being often more than a foot high and two or three inches diameter. It is called the "snow-plant" in California, because it appears before the snow has wholly melted and is most striking and beautiful when growing out of it. This plant is accurately represented in one of the pictures in the "North" gallery at Kew. On the sides of the rocky streams growing in fissures which are often under water, the large peltate saxifrage seems quite at home, although in our gardens it will grow and flower even in the driest situations. The fine shrubby Penstemon Newberryi also adorns the rocky margins of the streams, the beautiful Diplacus glutinosus of our greenhouses is a common wayside shrub, while the lovely blue Brodiæas and painted Calochorti or butterfly-tulips, are as common as our bluebells and poppies. The fine yellow Cypripedium montanum is occasionally found in the forest bogs, while in open ground near the "Big Tree" Hotel, exquisite little blue Nemophilas, yellow Mimulus, and a tall Echinospermum with flowers like a large forget-me-not, were very abundant. Among these and many other strange flowers one British species was found, often starring the ground under the giant trees with its delicate flowers. This was the little chickweed wintergreen (*Trientalis Europæa*), only differing from our native plant in the flowers being pale pink instead of white.

Even if we leave out of consideration the giant Sequoias,

the forests of the Sierra Nevada would stand pre-eminent for the beauty and grandeur of their pines, firs, and cedars. Three of these, the white pine, the red cedar, and the sugar-pine are, not unfrequently, more than six feet in diameter at five or six feet above the ground, whence the giant trunks taper very gradually upwards. One sugar-pine near the big-tree Hotel was found to be seven feet two inches diameter at five feet above the ground. A red cedar measured at the same height was seven feet diameter, and one of the white pines five feet nine inches. The height of the above-named sugar-pine was measured approximately by means of its shadow, and found to be 225 feet, and I was assured that one which had been cut down near the hotel was 252 feet high. The Douglas fir in the forests of British Columbia is said to surpass these dimensions considerably, being often ten feet or even twelve feet diameter, and near 300 feet high. Probably in no other part of the world than the west coast of North America is there such a magnificent group of trees as these; yet they are all far exceeded by two others inhabiting the same country, the two Sequoias--S. gigantea and S. sempervirens.

In the popular accounts of these trees it is usual to dwell only on the dimensions of the very largest known specimens, and sometimes even to exaggerate these. Even the smaller full-grown trees, however, are of grand dimensions, varying from 14 to 18 feet in diameter at six feet above the ground, and keeping nearly the same thickness for perhaps a hundred feet. In the south Calaveras grove, where there are more than a thousand trees, the exquisite beauty of the trunks is well displayed by the numerous specimens in perfect health and vigour. The bark of these trees, seen at a little distance, is of a bright orange brown tint, delicately mottled with darker shades, and with a curious silky or plush-like gloss, which gives them a richness of colour far beyond that of any other conifer. The tree which was cut down soon after the first discovery of the species, the stump of which is now covered with a pavilion, is 25 feet in diameter at six feet above the ground, but this is without the thick bark, which would bring it to 27 feet when alive. A considerable portion of this tree still lies where it fell, and at 130 feet from the base I found it to be still 12 1/2 feet in diameter (or 14 feet with the bark), while at the extremity of the last piece remaining, 215 feet from its base, it is six feet in diameter, or at least seven feet with the bark. The height of this tree when it was cut down is not recorded, but as one of the living trees is more than 360 feet high, it is probable that this giant was not much short of 400 feet.

The huge decayed trunk called "Father of the Forest," which has fallen perhaps a century or more, exhibits the grandest dimensions of any known tree. By measuring its remains, and allowing for the probable thickness of the bark, it seems to have been about 35 feet diameter near the ground, at 90 feet up 15 feet, and even at a height of 270 feet it was 9 feet diameter. It is within the hollow trunk of this tree that a man on horseback can ride--both man and horse being rather small; but the dimensions undoubtedly show that it was considerably larger than the "Pavilion tree," and that it carried its huge dimensions to a greater altitude; and although this does not prove it to have been much taller, yet it was in all probability more than 400 feet in height.

Very absurd statements are made to visitors as to the antiquity of these trees, three or four thousand years being usually given as their age. This is founded on the fact that while many of the large Sequoias are greatly damaged by fire the large pines and firs around them are quite uninjured. As many of these pines are assumed to be near a thousand years old, the epoch of the "great fire" is supposed to be earlier still, and as the Sequoias have not outgrown the fire-scars in all that time they are supposed to have then arrived at their full growth. But the simple explanation of these trees alone having suffered so much from fire is, that their bark is unusually thick, dry, soft, and fibrous, and it thus catches fire more easily and burns more readily and for a longer time than that of the other coniferæ. Forest fires occur continually, and the visible damage done to these trees has probably all occurred in the present century. Professor C. B. Bradley, of the University of California, has carefully counted the rings of annual growth on the stump of the "Pavilion tree," and found them to be 1,240; and after considering all that has been alleged as to the uncertainty of this mode of estimating the age of a tree, he believes that in the climate of California, in the zone of altitude where these trees grow, the seasons of growth and repose are so strongly marked that the number of annual rings gives an accurate result.

Other points that have been studied by Professor Bradley are, the reason why there are so few young trees in the groves, and what is the cause of the destruction of the old trees. To take the last point first, these noble trees seem to be singularly free from disease or from decay due to old age. All the trees that have been cut down are solid to the heart, and none of the standing trees show any indications of natural decay. The only apparent cause for their overthrow is the wind, and by noting the direction of a large number of fallen trees it is found that the great majority of them lie more or less towards the south. This is not the direction of the prevalent winds, but many of the tallest trees lean towards the south, owing to the increased growth of their topmost branches towards the sun. They are then acted upon by violent gales, which loosen their roots, and whatever the direction of the wind that finally overthrows them, they fall in the direction of the overbalancing top weight. The young trees grow spiry and perfectly upright, but so soon as they overtop the surrounding trees and get the full influence of the sun and wind, the highest branches grow out laterally, killing those beneath by their shade, and thus a dome-shaped top is produced. Taking into consideration the health and vigour of the largest trees, it seems probable that, under favourable conditions of shelter from violent winds and from a number of trees around them of nearly equal height, big trees might be produced far surpassing in height and bulk any that have yet been discovered. It is to be hoped that if any such are found to exist in the extensive groves of these trees to the south of those which are alone accessible to tourists, the Californian Government will take steps to reserve a considerable tract containing them, for the instruction and delight of future generations.

The scarcity of young sequoias strikes every visitor, the fact being that they are only to be found in certain favoured spots. These are, either where the loose *débris* of leaves and branches which covers the ground has been cleared away by fire, or on the spots where trees have been uprooted. Here the young trees grow in abundance and serve to replace those that fall. The explanation of this is, that during the long summer drought the loose surface *débris* is so dried up that the roots of the seedling sequoias perish before they can penetrate the earth beneath. They require to germinate on the soil itself, and this they are enabled to do when the earth is turned up by the fall of a tree, or where a fire has cleared off the *débris*. They also flourish under the shade of the huge fallen trunks in hollow places where moisture is preserved throughout the summer. Most of the other conifers of these forests, especially the pines, have much larger seeds than the sequoias, and the store of nourishment in these more bulky seeds enables the young plants to tide over the first summer's drought. It is clear, therefore, that there are no indications of natural decay in these forest giants. In every stage of their growth they are vigorous and healthy, and they have nothing to fear except from the destroying hand of man.

Destruction from this cause is, however, rapidly diminishing both the giant Sequoia and its near ally the noble redwood (*Sequoia sempervirens*) a tree which is more beautiful in foliage and in some other respects more remarkable than its brother species, while there is reason to believe that under favourable conditions it reaches an equally phenomenal size. It once covered almost all the coast ranges of central and northern California, but has been long since cleared away in the vicinity of San Francisco, and greatly diminished elsewhere. A grove is preserved for the benefit of tourists near Santa Cruz, the largest tree being 296 feet high, 29 feet diameter at the ground and 15 feet at six feet above it. Much larger trees, however, exist in the great forests of this tree in the northern part of the State, but these are rapidly being destroyed for the timber, which is so good and durable as to be in great demand. Hence Californians have a saying that the redwood is too good a tree to live. On the mountains a few miles east of the Bay of San Francisco, there are numbers of patches of young redwoods indicating where large trees have been felled, it being a peculiarity of this tree that it sends up vigorous young plants from the roots of old ones immediately around the base. Hence in the forests these trees often stand in groups arranged nearly in a circle, thus marking out the size of the huge trunks of their parents. It is from this quality that the tree has been named "sempervirens," or ever flourishing. Dr. Gibbons, of Alameda, who has explored all the remains of the redwood forests in the neighbourhood of Oakland, kindly took me to see the old burnt-out stump of the largest tree he had discovered. It is situated about 1,500 feet above the sea and is 34 feet in diameter at the ground. This is as large as the very largest specimens of the Sequoia gigantea, but it may have spread out more at the base and have been somewhat smaller above, though this is not a special characteristic of the species. Many other stumps were seen which were 20 and 30 feet in diameter, and all were surrounded with young trees of various sizes. The large tree is said to have been cut down forty years ago. It is, therefore, probable that, in the forests to the northward, redwood trees may exist equalling, if not surpassing, the "big trees" themselves.

I have now concluded a very brief and imperfect sketch of the more prominent aspects of North American vegetation, as seen during a single summer's travel across the continent. Many grand and beautiful scenes remain vividly painted on my memory; but if I were asked what most powerfully impressed me, as at once the grandest and most interesting of the many wonders of the western world, I should answer, without hesitation, that it was the two majestic trees some account of which I have just given, together with the magnificent and beautiful forests in the heart of which they are found. Neither the thundering waters of Niagara, nor the sublime precipices and cascades of Yosemite, nor the vast expanse of the prairies, nor the exquisite delight of the alpine flora of the Rocky Mountains--none of these seem to me so unique in their grandeur, so impressive in their display of the organic forces of nature, as the two magnificent "big trees" of California. Unfortunately these alone are within the power of man totally to destroy, as they have been already partially destroyed. Let us hope that the progress of true education will so develope the love and admiration of nature, that the possession of these altogether unequalled trees will be looked upon as a trust for all future generations, and that care will be taken, before it is too late, to preserve not only one or two small patches, but some more extensive tracts of forest, in which they may continue to flourish, in their fullest perfection and beauty, for thousands of years to come, as they have flourished in the past, in all probability for millions of years and over a far wider area.

Notes Appearing in the Original Work

    1. See *Biologia Centrali-Americana, Botany*, vol. i., pp.

lxvi.-lxvii.

2. American periodicals are full of accounts and illustrations of the poverty and hard lives of the small farmers. See, in *The Arena* of July, the article by Hamlin Garland, *A Prairie Heroine*.

www.ingramcontent.com/pod-product-compliance
Lightning Source LLC
Chambersburg PA
CBHW022343280326
41934CB00006B/754